Giuseppe Marino Urbani de Gheltof, Lady Layard

A Technical History of the Manufacture of Venetian Laces

Giuseppe Marino Urbani de Gheltof, Lady Layard

A Technical History of the Manufacture of Venetian Laces

ISBN/EAN: 9783744646543

Printed in Europe, USA, Canada, Australia, Japan

Cover: Foto ©berggeist007 / pixelio.de

More available books at **www.hansebooks.com**

A

TECHNICAL HISTORY

OF THE MANUFACTURE OF

VENETIAN LACES

(VENICE-BURANO)

BY

G. M. URBANI

DE GHELTOF

WITH PLATES

TRANSLATED BY

LADY LAYARD

VENICE
FERD. ONGANIA EDIT.
1882.

TO THE READER

he word Lace (Merletto) means, in its ordinary sense, a net work of linen, made by interlacing threads continuously without carrying them twice over the same ground. In the more general sense, it includes a trimming of any kind of net, or other similar work employed for ornament.

If we desired to arrive at the origin of this industry we should have to go very far back. The earliest garments worn by man were merely intended as a defence from the cold. The necessity of mending parts of them must have been felt long before the wish to adorn them. It is probable that the work of joining together the frayed edges, and of filling up those parts of them which may have been worn away, led to the idea of an ornament such as is placed on the edge of dresses, and is formed of threads more or less frequently, and more or less closely, crossed. Hence from the expression *racca* (*racconciare*, to mend) may have come the Italian word *ricami*.

The idea of a fringe must soon have suggested itself from the unravelling and fraying produced by wear on the edges of dresses; and this gave rise to the word *merletto*, and its equivalent in other languages, used as a generic name for any kind of ornament ending in many points, applied to the edges of garments, of handkerchiefs etc. etc.

This ornamental Industry arose, consequently, from the first necessities of life, and from the wear and tear of time and use upon the rude garments of primitive humanity. Under this aspect these poor garments may be considered as the oldest editions of books on Lace, although they have never found a place even in the most complete library.

Thus it would appear that the first step in lace-making was to embroider on a solid ground, which was then cut away and pierced. Next came the ornamentation of articles of dress by drawing out the threads of the texture, and by making edgings, or trimmings, by interweaving these threads with the needle without there being any ground - the ground being furnished by the work itself. Such was the origin of needle-made lace. Lace made with spindles, or lead bobbins (*a fuselli o a piombini*), now known as pillow lace, came from braiding or plaiting the frayed edges and loosened threads, and thus forming with them a network of different patterns, which could similarly serve for trimmings.

Under these two fundamental heads, *needle* and *pillow* lace, all the varieties of lace may be classed; such varieties only consisting in the different arrangement of the threads, and in the greater beauty of design.

The only practical way of treating of the development of any industry is to follow it through its natural changes, and so to proceed, little by little, from the simplest methods to the most complicated. Perhaps there may be exceptions to this rule; but they are more likely to occur when the art has reached a more advanced stage. They arise, however, from historical reasons to which it is not necessary to refer in an attempt to collect together the elements of one of the most elegant of industries.

As we must, therefore, restrict our explanations to Venetian laces, or to those which took their names from the city of the Lagunes, on account of the uses to which they were there so magnificently applied, we will endeavor to include in the two above mentioned categories — needle and pillow lace — the various sorts of laces which have been called after Venice.

The point-laces (*punti di merletti*) to which

all authors give the name of Venetian are the following :

Punto	a reticello	net lace
«	tagliato	cut lace
«	in aria	open lace
«	tagliato a fogliami	flowered lace
«	a groppi	knotted lace
«	a maglia quadra	darning or square netting
«	di Venetia	Venice point
«	di Burano	Burano point
«	tirato	drawn lace
«	burato	embroidered linen

It must be observed that those laces known as *a groppi, a maglia quadra, burato,* and *tirato* cannot be called true laces. The first is only a combination of small cords knotted together, very similar to the work of the *pasmenterie* makers; whilst the second, known as the *modano* of the Tuscans, is made by means of small pieces of wood, and is very like hand-made netting. *Burato* is an embroidery on coarse linen, and *punto tirato* is made by pulling out threads in a regular pattern, as in the *punto a reticello,* as we shall hereafter see. For this reason we will omit them from this notice confident that we shall thus do no injustice to the renown of the Venetian laces, as they are all simple works, except the *punto tagliato,* which contains the true rudiments of *needle* lace.

We have included in the Venetian laces pillow-lace, which, from ancient times, was also made in

Venice, and there reached such perfection that it may almost rank amongst them.

We will add that the different methods of lace-making which we have described, have been taken either from ancient works on lace, or from *viva voce* explanations given by the lace-makers themselves. We have, therefore, done nothing but preserve to posterity the technical details of a manufacture which has now been carried to the highest perfection.

We will conclude this preface by asking the reader to overlook any errors in this work, which would not have been written had it not been for the courtesy of a person from whom we have received the greatest assistance.

FFF

PUNTO TAGLIATO

(CUT-WORK)

The simplest of all the needle-made laces, which we have to describe, is Cut Lace — *Punto Tagliato*.

It is uncertain whether the white ornaments on the colored sleeves of some of the figures in Carpaccio's pictures, representing the Martyrdom of St. Ursula, are meant for this lace. There is, however, another picture of the same date, and by the same painter, in the Civic Museum at Venice, in which are depicted two ladies, in the rich costume of the XVth century, one of whom has on the edge of her dress a narrow border of white lace, exactly like that figured in Vercellio's *Corona*, and in many other books of the kind, and there called *punto tagliato*.

After the XVth, and during the whole of the XVIth century, *punto tagliato* remained in fashion. Matteo Pagan published, in 1558, the « Glory and

Honor of Cut Laces and Open Laces » (¹), and it was a Venetian, Frederico Vinciolo, who brought this Lace into great celebrity in France, printing at Paris, in the year 1587, the *Singuliers et nouveaux pourtraicts et ouvrages de lingerie,* wherein are specially given designs for *point couppé.*

The lace of which we are now treating, anterior to others in its origin, but contemporaneous with them in its early development, declined in public estimation as the other laces reached a perfection to which *punto tagliato* could never attain. It would, nevertheless, appear that it continued to be held in esteem during the first few years of the XVIIth century; for we find in the *Inventario dei drapamenti de lin* (Inventory of the linen garments) which formed part of the dowery of Cecilia de Mula, cloths *con lavorieri de ponte tagio grande* (with borders of broad cut-lace).

With regard to the methods used by our ancestors in making *punto tagliato,* it may be observed that, although they may vary in certain particulars which we will point out, they lead to the same results.

(1) La Gloria e l' honore dei ponti tagliati et ponti in aere.

PUNTO TAGLIATO

(CUT-WORK)

Materials

A round cushion to work upon (*balon*), sewing needles — large-headed pins — thread.

How to prepare the work

Draw on the linen the outline of the pattern to be worked, which is done by covering it with a piece of transferring paper, colored on one side only, and then tracing the required design; or else, prick out the outline, and afterwards, placing the paper on the linen, rub it over with coloring powder, which will leave the outline neatly indicated.

If you wish, for example, to copy the pattern of figure 1; having transferred the pattern to the

Fig. 1.

linen, and having fixed it with pins to the cushion,

2

trace out the design with thread. Then begin the lace work.

How to work

Work over the whole tracing with button-hole stitch, as in fig. 2, which shews the direction of the needle and thread. This is a very common stitch used in nearly all needle-work; but. nevertheless, it is as well to give a drawing of it, as the greater number of point-laces are composed of it.

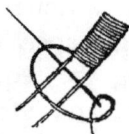

Fig. 2.

If you wish to ornament the small holes, or loops (*becchetti*), with purls, or *smerli*, be careful, when the button-hole stitch is done, to put through it, at the top, a large-headed pin; work across it several stitches according to the size of the purl required; take out the pin, and the stitch is finished. Then continue along the tracing as in figure 3.

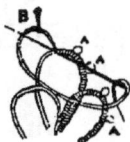

Fig. 3.

For greater variety, the button-hole stitch may enclose two parallel tacking threads, as in fig. 2.

With such very simple elements may be made any sort of *punto tagliato*. To finish the work, cut away the linen from the outline, and from the stitches which you may wish to leave open; thus making what is called *punto a giorno,* or open work. Take care not to destroy any part of the thread of which the stitches are made.

In this manner any one familiar with the ancient examples published in the numerous pattern-books which have been handed down to us, may easily execute what, towards the end of the XVIth century, was known as *invention de déesse*.

Another way of working this lace was to draw some threads across a little work-frame in geometrical pattern, which were then fastened down on a piece of linen ([1]) and covered with button-hole stitch, as in the preceding examples. Sometimes the work of interlacing the threads was done by carrying several of them in a radius from a knot placed in the centre, and afterwards working on them any kind of design, by means of the above mentioned button-hole stitch (*punto a festone*).

(1) Anciently called « *Quintain.* »

PUNTO A RETICELLO

(NET-LACE)

It is doubtful whether the particulars of the deed of division between Angela and Ippolita Sforza Visconti of Milan, of the year 1463, refer to works of *punto a reticello* by the words *radexelo, radexela, radexele, redicelle*. It is, however, certain that by *radexela* was often meant any kind of net, not of lace; but on the other hand, it is frequently used by authors to signify lace (*merletto*).

Thus Giacomo Franco, in his *Nuova inventione de diverse mostri* (New inventions of different patterns), printed in 1596, gives examples of laces which he calls *radixeli*, and which were a kind of ornament that might have been used in the 15th century.

Moreover, in the pictures of Gentile Bellini, in the Royal Institute of Fine Arts at Venice, there are represented several women with their necks and bosoms covered with the kind of lace which we are describing. Vercellio in the *Corona* also figures a lace very like Greek lace, which he calls *a reticello*. The wonderful perfection to which this lace had reached at the end of the 16th century is shewn by a portrait of Tensino prefixed to his works, printed in the last years of that century.

This lace greatly resembling *cut lace,* lost its value on account of the change of fashion; but two centuries later Francesca Bulgarini, who died in 1762, left, as is stated by Merli (*Origine delle trine* page 25), wonderful examples of it executed by herself, resembling the *reticello* of the Venetians.

As in cut lace, there were anciently two methods for working *punto a reticello.* We will describe them both, as we think it necessary to make our readers acquainted with them.

PUNTO A RETICELLO
(NET LACE)

First Method

Materials

The same as those used in *Punto tagliato.*

How to prepare the work

The design given in fig. 4 is an example of this Lace, taken from an ancient pattern book.

Fig. 4.

Take a piece of linen, on which mark out a number of small squares. Then pull out transversely, for instance, eight threads of the linen, so that nothing will be left but the perpendicular threads, as in A of figure 5.

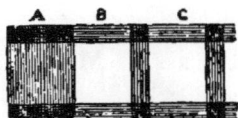

Fig. 5.

Count off four perpendicular threads, which will form, as it were, a lateral division between which, cut away, at their upper ends, other eight perpendicular threads, thus leaving an open space. This will form an equilateral square, as in B of figure 5. Continue in this way to treat all the four spaces; always remembering to leave the four threads which form the division and to cut away the next eight. If you wish to make an upper series of open squares, you must leave four transversal threads, which will thus form a division, as you have done in the lateral squares below.

How to work

When all the squares are made out, as in fig. 5, work over all the outlines with button-hole stitch, as in fig. 2. When this is done, the interior of the squares must be filled in with the patterns given in fig. 4. In order to do this, fix a thread in the upper corner to the left (fig. 6 A), if you are going to work the pattern A fig. 4; make one button-hole stitch; then a row of 3; then of 5 etc. etc. as in fig. 6, which is done by introducing the needle into the loop of the stitch of the row above, as in fig. 6.

Fig. 6.

When you have thus filled up half the square, finish the work by cutting the thread.

The small purls, indicated in A of fig. 4, must be done as you work, and when you are making the last row of the button-hole stitch add two or three stitches where you wish to have the edging, and then continue the button-hole stitch.

Fig. 7.

The design B in fig. 4 is done by drawing several parallel threads from A to B of fig. 8, and fastening them at those points.

If, for example, there are six threads, divide them into two groups by covering three with but-

ton-hole stitch and afterwards adding the *purls,* as
in fig. 7.

Fig. 8.

The pattern C can be worked, as in fig. 9, by
drawing two threads across from A to B and from
C to D, which will cross each other; then adding
a circle of thread which is fixed to the others with
button-hole stitch, and with threads drawn from
the sides; these are also to be covered with the
same stitch.

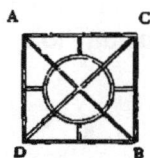

Fig. 9.

The border, or lace edging, is done in a different way, by changing the arrangement of the threads,
but generally by drawing across a semicircle of thread
from A to B, as in fig. 10, which is fastened at both
ends, and afterwards covered with button-hole stitch,

and by filling up the open spaces with threads ar-

Fig. 10.

ranged according to fancy, which are also to be covered with button-hole stitch.

SECOND METHOD

Materials.

Paper on which is drawn the design to be
worked; thread of different sizes, such as is used
in making other kinds of lace.

How to prepare the work.

To work the design given in fig. 4, place on
the paper, on the outlines of the pattern, a much

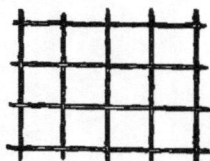

Fig. 11.

coarser thread than is used for the rest of the work, and fix it with a stitch. You will thus have made a net work, as in fig. 11.

All the threads of this net-work must then be covered with button-hole stitch. The patterns in the figures 8, 9 and 10, as also the border, or lace-edging, are made as explained in the preceding method. Lastly, detach the work from the paper.

PUNTO IN ARIA

(OPEN-LACE)

In the *Opera Nuova* of Giovanni Antonio Tagliente, printed at Venice in 1530, there is described, amongst other laces, that called *in aere*,
or open lace.

It is not certain whether Tagliente intended
to allude to a special stitch used in needlework, or
to that combination of stitches which afterwards
came to be known by the same name.

The works entitled « *Le Pompe* » of 1557, and
« *La Gloria et l'honore dei ponti tagliati et ponti
in aere* » (*The Glory and honor of cut and
open laces.*) of the year 1558, are the first which
give examples of the true *punto in aria*, subsequently
so greatly esteemed in Venice and elsewhere.

The inventory of the furniture of Gio: Batt:
Valier, Bishop of Cividale di Belluno, made in 1598,
mentions five pieces of bed linen of needle-
worked point . . . *ancient works,* and pillow

cases of the same lace, besides ten napkins (*for-nimenti di tovaglioli*) of similar work equally old.

The *punto in aere* was also included in the sumptuary laws of the Republic. In the years 1616, 1633 and 1634 the *Provveditori alle Pompe* proscribed its use in Venice under a penalty of 200 ducats for each offence. But fashion retained it in spite of the laws.

Towards the end of the 16th century, Cesare Vecellio published his « *Corona* », which was reprinted in the years 1600, 1620 and 1625, and the Venetian lace-makers continued to work these laces which, in 1694, obtained the praises of Zunica in his « *Calamita d'Europa.* » He calls them *beautiful laces which are made with the needle, and woven with such fine threads that they are worthy of the name of punto in aria.*

The collar destined for the Coronation of Louis XIV worked by Venetian women, who employed in making it their beautiful fair hair, was of *punto in aria.*

This lace remained in fashion in the 18th century. Even Joseph II ordered, of a merchant of Venice, a complete set of it at the price of 77777,78 Italian lire. However, at that time the Venetian ladies despised point-lace, preferring the laces of Flanders which, although very wonderful, were cartainly not equal in renown to Venetian *Punto in Aria.*

PUNTO IN ARIA

(OPEN LACE)

Materials.

The same as those used in the preceding laces.

How to prepare the work.

Draw on a long strip of paper, of the width
of the work to be executed, the outlines of the pattern. Be careful to fasten under this paper a piece
of linen, in order to strengthen it.

If, for example, you wish to work the design
in fig. 12, first outline the work, and the places

Fig. 12.

which are required to be in the highest relief, with a coarse thread, as in *punto a reticello*.

How to work.

Cover the thread of the outline of any part of the pattern with button-hole stitch.

The spaces within the outlines must then be filled in with button-hole stitch, or may be worked with net-work (*reticelli*), according to any of the designs of which we give examples.

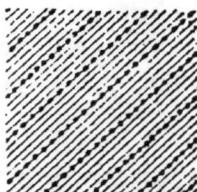

Fig. 13.

To execute this design (Fig. 13), you must first make three or four rows of button-hole stitch ; then begin another row, which is to be ornament-ed with small open work. Take care, at every open-ing you wish to make, to miss a stitch, and to put your needle into the next one.

Having finished this row, do another of simple button-hole stitch, which will include the stitches left open.

Remember, while you are working the open stitches (*in aria*), to stick a pin into the opening, or to run in a thread, in order to keep it clear. Then begin again the rows of plain button-hole stitch, alternating them with the above mentioned open work.

The same rules apply to the working of the following examples (Fig. 14, 15, 16 and 17)

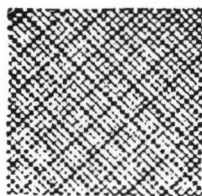

Fig. 14.

Fig. 15.

Fig. 16.

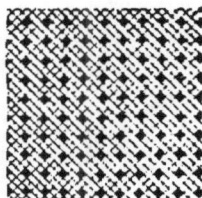

Fig. 17.

In order to join the different parts of the pattern together, you must unite them by many threads, covering them with button-hole stitch.

The small purls described in *punto tagliato* may also be added, as an ornament to this lace.

4

(I

PUNTO TAGLIATO A FOGLIAMI

(FLOWERED LACE)

The lace called *Punto tagliato a fogliami,* or flowered lace, acquired a greater renown than any other made at Venice, on account of its elegant and graceful designs. Nevertheless, its origin must be referred to a period in which all the arts were in the greatest decline. In the second half of the 17th century artists produced excellent effects by introducing this lace in their portraits of illustrious Venetians. It may, therefore, be presumed that it was only in use from that time.

We may mention some engravings of the Doge Francesco Morosin;, whose wonderful laces, still jealously preserved in the palace of his family, resemble those represented in his portraits. Nor can we omit to notice the laces of the Dogaressa Quirini Valier , in a painting existing in the Civic Museum at Venice, and those in many other portraits of Venetian magistrates who flourished from the end of the 17th to the first half of the 18th century. It may also be confidently asserted that *punto tagliato a fogliame* was amongst the principal ornaments of the churches and 'of the dresses of the ecclesiastics of that time.

One branch of this lace was called *punta rosa* (rose - point ,) from its resemblance to the flowers from which it takes its name. The collars of the magistrates of the old Republic were richly co-vered with this splendid rose-point, worked by Ve-netian women. Their rivals of the present day are not inferior to them in producing this lace in de-signs of the greatest variety.

PUNTO TAGLIATO A FOGLIAMI

(FLOWERED LACE)

———

Materials.

The same as those used in the preceding lace.

How to prepare the work.

Draw the outline of the pattern on slightly tinted paper; afterwards fasten under it a thicker paper attached to a piece of linen, in order to strengthen it.

To execute figure 18; mark out the out-

Fig. 18.

lines with a coarse thread; then commence by filling in all the spaces with button-hole stitch, varying the designs at pleasure, as in the patterns given in figures 13, 14, 15, 16 and 17, in order to imitate the forms and texture of flowers and leaves.

Having done this, cover the outline of those parts which you wish to be in the highest relief with several threads, and then work them over with button-hole stitch, ornamenting them with purls, as explained in the preceding laces.

Then cover all the other outlines of the work with the same stitch.

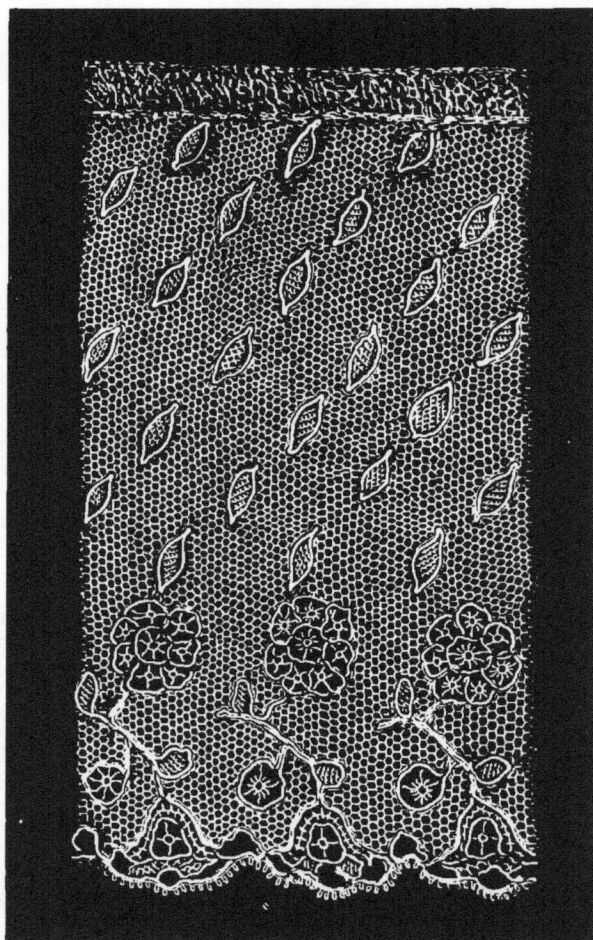

PUNTO DI BURANO

(BURANO POINT)

The origin of this lace is as uncertain as that of other laces. The first traces of it are found in the 18th century.

The *Gazzetta Veneta* of 1792 first noticed *punto di Burano*, in which there had been a large trade even in ancient times. In the portrait of Alvise Pisani, engraved en 1793 by Bartolozzi, is the only example of this lace to be found in the rich collection of engraved portraits, from the 17th to the 19th century, preserved in the Civic Museum of Venice.

Moschini in his *Itineraire*, printed in 1819, and the last writers who make mention of Burano, speak of the lace worked there by its handsome women. Some pupils of its institution for girls were a few years ago rewarded by the Royal Venetian Institute for their excellent samples of this industry.

But the boast of the lace-makers of Burano is
that their work has been greatly appreciated in
the present day. The school there, founded by the
generous efforts of some noble ladies and praise-
worthy citizens, is a new source of pride to the
little island of the Lagunes, and to Venice whose
fate it shares (1).

(1) See Appendix.

PUNTO DI BURANO
(BURANO POINT)

Materials.

A common round cushion for working: needles and pins of different sizes: a small wooden cylinder (1): thread of various sizes.

How to prepare the work.

Draw the pattern in simple outline on coloured paper: fasten under it a piece of linen of the same size, as in working other laces.

Fig. 19.

In order to work this design (fig 19), fasten

(1) Called in the Burano dialect *Morello.*

the paper having the outline on the cushion with pins; then put under it the little wooden cylinder, as shewn in A of fig. 20.

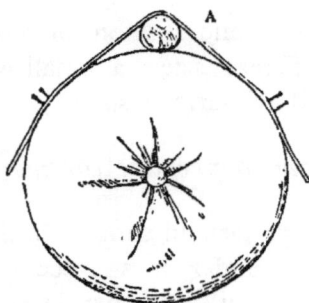

Fig. 20.

Outline the pattern with a coarser thread than that to be used for the ground.

Burano lace is divided into three distinct parts — the ground, the flowers, and the border.

The first thing to be done is to make the ground; to do which, fasten a thread at the upper part to the left of the outline, as in figure 21; then carry the thread from the top to the bottam If in doing this you come to other outlines, which prevent the drawing of the thread in a straight line, as in A of fig 21, fix it at the outline with a simple stitch, and continue on the other side. When the thread A reaches B (in fig. 21), fasten it, and make the first stitch of the net work, passing the

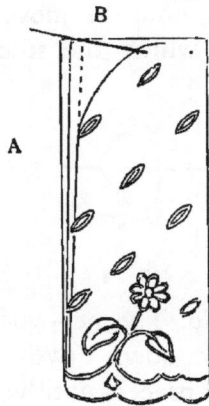

Fig. 21.

needle over the second thread, and under the first, as in fig. 22.

Fig. 22.

Then hold the thread tight between your fore-finger and thumb (fig. 23), and insert the needle under the second thread, as in fig. 23, making a

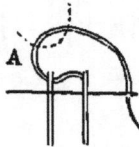

Fig. 23.

stitch, but always continuing to hold the thread
A with your fingers, until the movement shown in
fig. 24 is made, by which the stitch is closed by

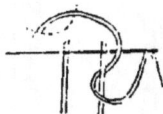

Fig. 24.

pulling up the needle. The net-work is thus com-
pleted. Then carry down two more threads,
and follow the same process until you have finished
the whole of the ground.

The spaces within the outlines of the flowers,
and of the border, may be filled in, according to
fancy, with the patterns given in fig. 13, 14, 15, 16,
and 17, for the *punto in aria*.

Lastly, the outlines of the flowers are to be
worked over with button-hole, or any other stitch.

PUNTO DI VENEZIA

(VENICE POINT)

The first traces of the manufacture of this lace are to be found in the beginning of the 17th century. The ample dresses of that date required trimmings adapted to the great richness of the velvets and brocades then in use, and the designers soon employed their fancy in inventing beautiful patterns.

At soon as this lace had been introduced into France it obtained snch celebrity that the wise Colbert, knowing the great benefits which might be reaped from its manufacture, brought over to France, at great expense, several Venetian lace-workers to found a school.

Punto di Venezia thus transplanted to a foreign soil, with a few modifications, came to be called « French point » and obtained the greatest celebrity

In the meanwhile, the Venetian factories continued to supply as much of it as Venetian fashion required, and even to flourish.

The following curious document will serve to show how much Venetian Point was used in our churches, and the great sums which were spent on it (1).

« 1769 : Feb ; 14. Cost of materials and manufacture of fine lace of Venetian Point, for two surplices for the Venerable *Scuola di S. Maria della Carità.*

For thread to work the lace . . , . L. 280.
For designs . . . . « 80.
For the make of 14 braccia of wide lace at
    L. 88 the braccio, and 6 braccia of
    narrow lace at L. 44 . . . . . « 1496.
For edging 20 braccia . . . . . . « 10.
For common linen, and other small expenses « 30.

L. 1896.

I, the undersigned, have received on account of the above materials and manufacture from the Rev. Sig. D. Giambattista Scioppalalba, Chaplain to the Venerable Scucla above mentioned, in several payments nine hundred and twenty eight lire and three soldi, that is L. 928.3.

MARIA TAGLIAPIETRA. »

(1) Royal Archives of Venice — Scuola della Carità. — Approvazione di Parti.

The making of this lace is so laborious as to render its manufacture very painful; so much so that its production and the trade in it, which existed in former times, have been greatly diminished.

———————

PUNTO DI VENEZIA

(VENICE POINT)

Materials.

Those used for *Punto in aria.*

How to prepare the work.

The same rules are applicable which serve for *Punto in aria,* as to the making and covering of the outlines.

The design of fig. 25 represents an ancient example of Venetian Point.

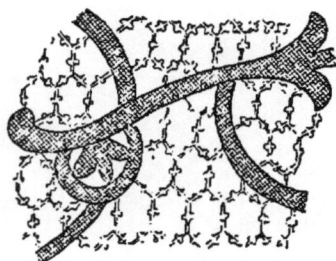

Fig. 25.

Note that the spaces between the outlines may either be filled with button-hole stitch, or with *punto a cordella,* darning stitch, This is done by drawing threads across from one side of the outline to the other, and afterwards passing the needle under and over them, thus making a kind of woven texture, as shown in fig. 26.

Fig. 26.

The ground of the pattern is made by small threads covered with button-hole stitch, to which is added, in working, the small purls already described in the *punto a reticello*.

MERLETTO A FUSELLI

(PILLOW LACE)

In the inventory of the expenses of the Coronation of Richard III of England it is stated, that the Queen wore a mantle of cloth of gold, with a trimming of white and gold silk from Venice.

It would appear, from the history of this industry, that these ornaments were the first laces made with bobbins.

But the first mention of this sort of lace is only found in the year 1596 in the *Nuova invenzione* published by Giacomo Franco, which gives two patterns of *ponto in Tombola* (lace made with bobbins) for sheets and handkerchiefs.

Both Vinciolo and Parrasoli, who published their works in the first years of the 17th century, give examples of *merletti a piombini* (lead-bobbin-laces), which, in fact, correspond with *merletto a fuselli* (bobbin-lace), and which were very much prized in Venice, as would appear by the many specimens of old lace of this description still to be found there.

The manufacture of pillow lace is now widely spread, especially in the islands of the Venetian estuary, on account of its being so easily made, and of its low price, which permits of a large sale. The women of Pellestrina and Chioggia, who work for the Society for the manufacture of lace, are able to supply a very large quantity of this beautiful production.

———————

MERLETTI A FUSELLI

(BOBBIN LACE)

Materials.

A common work cushion; a quantity of small wooden bobbins; a number of metal pins; and any sort of linen, cotton, silk or woollen thread, are the materials required for making pillow lace. (Fig. 27 and 28).

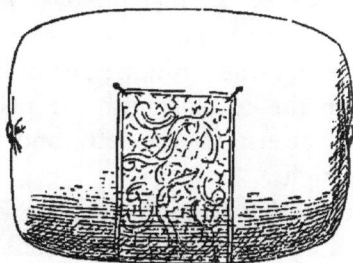

Fig. 27.

Fig. 28.

How to prepare the work.

Draw on a piece of strong paper the outlines of a pattern to be selected from any piece of lace

of which a copy is desired. The part to be covered
with the work is now generally coloured yellow, and
the spaces not to be filled in are left white. Place
under the drawn pattern two or three pieces of the
same paper, in order to economise time and labour
in making repeats of the pattern and to continue the
work, when the first piece of the design is finished, by
fixing them to the cushion; then pro ced to prick
the design, an operation which consists in piercing
those parts in which any figure, either a square
triangle or hexagon, is to be executed. These pins
will be afterwards of the greatest use.

Having finished this, take the design and the
pricked pieces off the cushion; one of them is then
fastened on to the cushion with the plain or straight
edge of the pattern to the left, and the bordered
edge to the right.

Then take the cushion on your knees, or else
place it on a stool, with its two ends so arranged
that the bordered edge shall be to the right, and
the plain edge to the left.

Have ready as many bobbins, wound with
thread, at the part A in fig. 28, as there are stitches
required in the design. Then take a large pin and
stick it into the cushion at a little distance from
the pattern, and fasten to it the end of a thread
of one of the bobbins, by twisting it round two
or three times from left to right; at the fourth
round fix this thread by a loop.

Then pull this loop tight, and the thread will be

fastened to the pin, and the bobbin will remain hanging to it. Unwind from the reel of the bobbin as much thread as is required for working, and prevent it from unwinding further by twisting the thread two or three times on the lower head of the bobbin, either from the right or the left, finishing with a loop. Put on to the same pin as many bobbins as it will hold, and then move it to the top of the pillow, at a little distance from the pattern. Load a second pin with bobbins, and place it in a horizontal line with the first; then a third etc. etc. until all the bobbins are in their places (fig. 29).

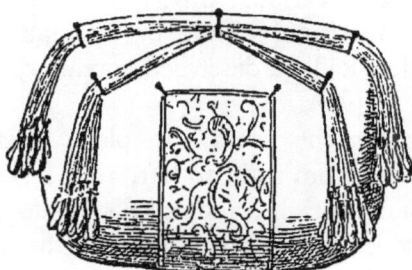

Fig. 29.

The pattern of lace to be copied is placed behind the row of pins to which the bobbins are hung.

How to work.

Arrange four bobbins in a row; throw the second over the first, the fourth over the third, the second over the third. Begin again, putting the second over the first, the fourth over the third, the second over the third; and thus what is called a *plait of eight* will be made. If, instead of the bobbins being used in twos, they are used one by one, a *plait of two* will be made.

When the plaits are reduced to the same length they must be pulled vertically and parallel the one to the other, and a pin fixed at the angle which the threads form at each of their extremities, leaving the first and second bobbins to the right, and the third and fourth to the left, of the pin which holds them apart.

There are various ways of fixing the plait; either by an ordinary knot, or by making a *cast stitch*. which will be described hereafter, or else by an ordinary stitch.

In making the braid, or plait, resume in a contrary sense from right to left, after having worked from left to right, and take care to leave two bobbins, which will serve to enclose the pins; then make an ordinary stitch.

From the ordinary stitch you may proceed to the weaving, or canvass, stitch. The weaving stitch begins where the ordinary stitch finishes; therefore, if it is to the left, leave the two first bobbins, take the four following ones, twist them by twos, that is to say, pass them from over under, and from under over the threads to which they are attached, numbering them from left to right, 1, 2, 3 and 4;

put the first over the third, the second over the first, the fourth over the third, and the second over the third, and the canvass stitch is complete.

To continue the work do not twist the threads, but, of the four bobbins you have been using, leave the two furthest to the left; take the two which remain, and to them add the two nearest, going, therefore, from left to right; put the second over the third and continue as before. The first movement is the only one which differs from the usual rule; in the first case the *first* is put over the third, and in this instance the *second* is put over the third. .

In order to work the *corona*, or finishing edge, of the lace, twist the two bobbins as you like; fix a pin at the point where they are twisted; pass them over this pin, and wind round it, from right to left, the twisted threads of the two bobbins. Then take of these two bobbins the one which is to the left, and pass its thread over the pin, returning from left to right over the head of this pin.

This last is only done to close and tighten the work, because when it is tightened the bobbin is to be put back as it was before.

When you have continued in this way until you get from the right to the left, four bobbins will remain, Separate these last four bobbins with one pin on one side and two on the other; twist the two bobbins of one side together, and similarly the two others, as you may wish, and end with a simple stitch, throwing the second over the third, the fourth over the third, the first over the second, and the second over the third, and so on consecutively.

You can close the work of the canvass stitch with net-work in the following manner :

Leave the two bobbins, and twist the two next threads. With these two bobbins, and the next two which are not twisted, make a stitch. Take the two last bobbins of this stitch and the two next; twist them by twos as they come, and make a stitch. With the four last of the sixteen, which are twisted in twos, work round a pin, and so on backwards and forwards. With the last four make a stitch without a twist, and then the edging to close the net.

If you wish to make an open-work ground, leave the two first bobbins from left to right, and work with the following four; then make a stitch; twist the first two of these four; keep hold of the last two; take the next two; twist all four by twos, and make a stitch; then place a pin betwen the last four, a little below the former pins; take the last four of the first eight; twist them by twos, and make a stitch; then separate them with a pin, and so on. When you come to the last four, they are not to be twisted, but make a stitch, then add the *corona*, or edging, and lastly another stitch.

The cast stitch is worked by taking the first four bobbins to the right; holding them by twos, making a stitch, twisting them again to make another stitch; then holding the next four bobbins, work with them as in the four preceding ones, finally, take the four next bobbins, and go through the same process, as with those that precede them and so on.

You may close the cast stitch by leaving the two first bobbins on the left; twisting the four following by twos, making a stitch, and fixing them with a pin. Next take the two preceding and the two following bobbins and twist them by twos, making a stitch and fixing it with a pin, and continue in this way until you come to the last six. Then work with the four penultimate bobbins, twisting the first two of them, and make a stitch. To finish off the ends, take the last four, twist them by twos, and make a stitch.

APPENDIX

The Burano School for lace was established about eight years ago in the Island of that name which, in the XVI. and XVII. Centuries, was one of the principal seats of the celebrated, Lace manufacture of the Venetian Provinces. The formation of the School, and the revival of the Art in Burano, arose out of the great distress which, in 1872, overtook its inhabitants, consequent on the extraordinary severity of the winter of that year rendering impossible the pursuit of fishing, upon which the population of the Island, consisting for the most part of fishermen, depend. So great was the distress at that time that they and their families were reduced to a state bordering on starvation, and for their relief money contributions were made by all classes in Italy, including the Pope and the King. This charitable movement resulted in the raising of a fund which sufficed to relieve the immediate distress, and to leave a surplus applicable to the establishment of a local industry likely to increase permanently the resources of the Burano population.

Unfortunately the industry fixed upon, that of the making of fishermen's nets, gave no practical result, the fishermen being too poor to purchase the nets. It was then that, on the suggestion of Signor Fambri, an effort was made to revive the ancient industry of Lace Making. Princess Chigi-Giovanelli and Countess Andriana Marcello were asked to interest themselves in and to patronise a school for this purpose, and to this application those ladies yielded a ready assent. At a later period Queen Marguerite graciously consented to become (and still is) the President of the Institution.

When Countess Marcello (who from that time has been the life and soul of the undertaking) began to occupy herself with establishing the School, she found an old woman in Burano, Cencia Scarpanile, who preserved the traditions of the Art of Lace Making, and continued, despite her seventy years and upwards, to make « *Burano Point.* » As she, however, did not understand the method of teaching, the assistance was secured of the Signora Anna Bellorio d'Este, a very skilful and intelligent woman, for some time mistress of the girls school at Burano, who, in her leisure hours, took lessons in Lace Making of Cencia Scarpanile and imparted her knowledge to eight pupils, who, in consideration of a small payment, were induced to learn to make Lace.

As the number of scholars increased, Signora Bellorio occupied herself exclusively in teaching Lace Making, which she has continued to do with surprising results. Under her tuition, the School, which in 1872 consisted of the eight pupils (who, as before mentioned, received a daily payment to induce them to attend), now numbers 320 workers, paid, not by the day, but according to the work each performs. In this way they are equitably dealt with, their gains depending on their individual skill and industry.

In Burano everything is extremely cheap, and a humble abode capable of accommodating a small family may be bought for from 600 to 1,000 Italian lire, and it is not a rare occurrence to find a young girl saving her earnings in the Lace School, in order to purchase her little dwelling that she may take it as a dower to her husband. Nearly all the young men of Burano seek their wives from among the Lace workwomen, and the Parish Priest reported last year facts which showed conclusively that the moral condition of the Island, consequent on the establishment of the Lace School, has improved in a very striking degree.

The Lace made in this School is no longer exclusively confined to Burano Point; but Laces of any design or model are now undertaken and especially those known under the following names:—

1. Point de Burano.
2. Point d' Alençon.
3. Ancient Point de Bruxelles.
4. Point d' Argentan.
5. Rose Point de Venise.
6. Italian « Punto in Aria. »
7. Italian « Punto tagliato a fogliami. »
8. Point d' Angleterre.

The following facts relating to the *modus operandi* in the School, may be of interest. '

In order the better to carry out the character of the different Laces, the more apt and intelligent of those pupils whose task it is to trace out in thread the design to be worked, have the advantage of being educated by means of Drawing lessons from professional artists.

The 320 workwomen now employed are divided into seven sections in order that each may be engaged upon one sort of work, and, as far as possible, one class of Lace. By this method each worker becomes thoroughly proficient in her own special department, executes her task with greater facility, earns consequently more, and the School on its part gets the work done better and cheaper (although of course cheapness must always be very relative).

The work of *the first section* is confined to preparing the outline of the ,design, and numbers 15 workers.

The second section employs 60 workers, who execute with the needle the ground work of little square holes which serves for the « *Burano Point* », and for part of the « *Point d' Alençon* », as also for. « *Ancien Point de Bruxelles* ».

The third section consists of 25 workers, who are trained to execute the ground-work of small round holes for the « *Point d' Alençon* » and the « *Point d'Argentan* ».

The fourth section consists of 100 workers, who execute simple guipure, or the stitches that fill up the flowers.

The fifth section employs 80 workers, who join the different pieces of the Laces together, and work the high relief to the flowers. In order to be qualified for this section the pupil must have attainded proficiency in every section of Lace Making, and it is to the workers in this and in the first section that the Drawing lessons are given.

The sixth section consists of about 10 workers, whose business it is to detach the Lace when finished from the pattern paper, clean it, and prepare it for sale.

The seventh section comprises those workers who, being married and therefore often occupied with family duties, cannot be subject to the same work hours as the single women.

VENEZIA. STAB. KIRCHMAYR E SCOZZI.

CATALOGUE

OF A COMPLETE COLLECTION OF

ANCIENT WORKS

ON

$\mathcal{V} E \mathcal{N} I C E \ L \mathcal{A} C E$ [1]

I. VECELLIO CESARE. — The Crown of noble and virtuous ladies, in which are shewn in 119 designs all manner of patterns of cut-laces, of edgings, etc. Fac-simile of the original edition of 1600. Venice. 1876 Fr. 80.—

II. FRANCO G. — A number of newly invented designs for giupure and net-lace; 24 patterns for laces etc. Fac-simile of the original edition of 1596. Venice 1877. « 20.—

III. III *bis*. The ROMAN LUCRETIA; A noble ornament for gentle matrons, wherein are contained collars, and borders of infinite beauty; 20 engravings in 4to. Fac-simile of the original edition of 1620. Venice 1876. « 60.—

IV. PAGAN MATT. — A good example of the laudable desire of noble-minded ladies to learn the art of making giupure laces ; with 31 engravings. Fac-simile of the original edition of 1550. Venice 1878. « 30.—

V E N I C E

L A C E

[1] The titles of these works are translated literally from the Italian.

V. Zoppino. (known as Nicolò Aristotile). — A general collection of beautiful ancient and modern embroideries, in which either man or woman may worthily exercise rare skill with the needle etc.; with 52 engravings. Fac-simile of the original edition of 1537. Venice 1877. Fr. 30.—

VI. Vavassore And: (called Guadagnini) — New universal work, entitled the Crown of Embroideries, in which worthy ladies and maidens will find various patterns for making collars of shifts, covers of cushions, silk coifs of many kinds, and a large number of works for embroiderers; with 40 engravings. Facsimile of the original edition of 1546 (!) Venice 1878. « 30.—

VII. Ostaus John (The true perfection etc). — A most delightful way of occupying your daughters with work, such as the chaste Roman Lucretia gave her maidens, and upon which they were found employed with her by Tarquin and her husband Collatinus, as described in the first book of the Decades of Livy; with 40 engravings. Fac-simile of the original edition of 1567. Venice 1878. « 30.—

VIII. Pagan Matio. — A new work composed by Domenico da Seva, called the Franciosino, in which noble and elegant young ladies are taught every kind of stitch, to sew, to embroider, and to do all such beautiful work as is fit for virtuous maidens, and for those who delight in making fancy-work with their own hands; and which is also very useful to weavers who are in the habit of working in silk. It comprises 12 small sheets, one within the other, and is signed AI-AXII; with an illustrated frontispiece and 46 engravings. Fac-simile of the original edition of 1546. Of this first edition, which is quite unknown to bibliophiles who have specially dealt with this subject, no other copy is believed to exist. Venice 1878 « 40.—

IX. Paganino Alex. (Burato) — First book of embroideries, by which, in various ways, the method and manner of embroidering are taught — a thing which has never yet been done, nor shewn, and which the reader will learn on turning the page, *In fine:* P. Alex. Pag. Benacenses, F. Bena V. V. *Without date, but certainly one of the first and most precious books of this kind;* with 28 engravings. Fac-simile of the original edition of the 15th century (!) Venice 1878. « 40.—

MONOGRAPHES

URBANI DE GHELTOF G. M. — The laces of Venice (History of), with the bibliography of the Venetian editions of works of instruction on lace; in 12° of 60 pages. Venice 1876.   Fr.   4.—

CERESOLE VICTOR — The origin of Venice Lace; and of the school for Burano point in 1878. (Universal Exhibition of Paris 1878. Group IV. class 36) not on sale.   «   —

M.rs BURY PALLISER. — A history of Lace. Third edition in 8.vo pag. 454. London 1875   «   45.—

URBANI DE GHELTOF G. M. — Treatise on the technical history of the manufacture of Venetian Laces (Venice-Burano) In 12° with 7 engravings in fac-simile and 27 vignettes. Venice, 1878.   «   15.—

VENEZIA. STAB. KIRCHMAYR E SCOZZI.